“Danger! Danger!” Barney shouted. “A large donkey is stuck in the chimney. I never saw such a strange sight! I’ll page the other animals for help.”

“How strange!” said Sophy Gopher.

“Only in America,” added Philip Q. Monkey.

“Get the ladder,” commanded Shelley Turkey. Then Shelley Turkey went into action.

The animals went up the ladder. Ginger Goat got up on Barney. Sophy and Philip got up on Ginger. And Shelley stood at the top.

“Change!” Shelley shouted out to the others. “All animals change places right now!” They changed places for her.

But the animals still could not reach the donkey.

"Only in America," said Philip. He liked that strange phrase.

"I paged that stranger for help," said Barney.

"Do you have a fireplace?" asked the stranger.

"In the study," Shelley answered.

“Count to seven,” the stranger told them. “Then start a large fire.” And up, up, up he went.

“Thanks!” the large donkey yelled.

“What now?” asked Barney.

“We animals can toast these in the study,” said Shelley.

The seven animals never had such a good time.